Calculations in Chemistry

Book 2

Concentration and Acid-Alkali Titrations

By T. Everitt

Calculations in Chemistry
A Series of Books for A-Level Students

Chemistry isn't all about calculations. They are not the be-all and end-all of A-level Chemistry either but it helps to be good at them. By definition this isn't a whole course textbook and not one full of exciting new discoveries or colourful content. Nor is it specific to a particular examination board.

The aim of this book is to provide examples to practise on in addition to those in your course textbook, which often do not contain many examples of questions and problems to test your understanding.

Some calculations are relatively simple, others more involved with multiple steps. The aim of this book is to guide you through examples and then give you questions to apply your understanding to. These begin simply but increase in complexity as you progress. Some also drop in the occasional non-calculation parts to keep you grounded in the reality of examination questions. Fully explained answers are included at the back of the book.

There are some key formulae that you **HAVE** to be able to recall. They enable you to answer questions speedily even if the formula is given somewhere in the examination paper and are essential if it is not. This recall also gives you the confidence that you are "doing it right!"

I hope that you find this book useful.

MrEV

Calculations In Chemistry Book Series

Book 1 Moles for A-level Chemistry

Book 2 Concentration and Acid–Alkali Titrations

Book 3 Analysis and Purity Calculations

Book 4 Enthalpy Calculations

Book 5 Entropy and Gibbs Free Energy

Book 6 Acids, Alkalis and Buffers

Book 7 Rate Equations and Equilibria

Book 8 Electrochemical Cells and Redox Equilibria

Concentration and Acid–Alkali Titrations

Contents

Part 1 – Concentration Calculations

Moles calculated from solutions is such a common calculation in chemistry you need to learn the equation. It doesn't easily work as an equation triangle unless you convert the volume into dm^3.

You can think of the moles when using a particular volume as "what fraction of the 1000 cm^3 solution have you taken from the 1000" multiplied by the concentration.

Equation to Learn:

$$\text{Moles} = \text{concentration} \times \frac{\text{volume}}{1000} \quad \textbf{OR} \quad \text{Moles} = \frac{C \times V}{1000}$$

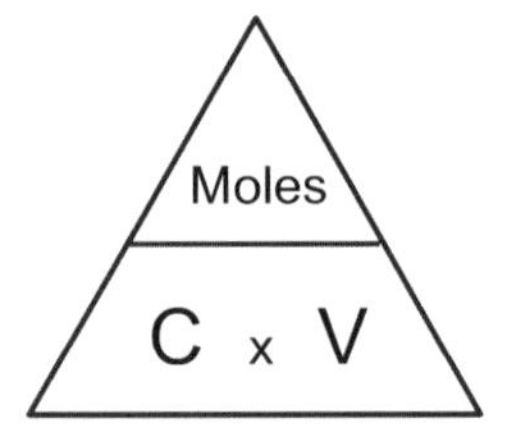

However, if you convert the volume from cm^3 to dm^3 by dividing V by 1000 then the equation works in a familiar triangle.

<u>Expected prior knowledge – you should be able to:</u>

- Be able to construct correct formulae
- Be able to construct balanced chemical equations
- To be able to rearrange simple mathematical equations
- Be able to calculate formula mass
- To be able to calculate moles of substance from a given mass
- To be able to calculate moles from concentration
- To be able to select appropriate numbers of significant figures for an answer from the data within the question

Example 1

a) How many moles are there in 40.0 cm^3 of a 0.0400 mol dm^{-3} $K_2SO_{4(aq)}$?

b) What is the concentration of sulfate ions in this solution?

c) What is the concentration of potassium ions in this solution?

Part a)

0.0400 x 40.0 / 1000 = **0.00160 moles.** [Note the M_r of potassium sulfate is not used in this question.]

Part b)

Potassium sulfate in solution ionises according to the equation below.

$$K_2SO_{4(aq)} \rightarrow SO_4{}^{2-}{}_{(aq)} + 2\,K^+{}_{(aq)}$$

If the concentration of $K_2SO_{4(aq)}$ is 0.0400 mol dm^{-3} then there is an equal concentration of sulfate ions (1 sulfate ion in the formula) so the concentration of sulfate ions = **0.0400 mol dm^{-3}**.

Part c)

From the equation in part b), there are twice as many potassium ions as the moles of $K_2SO_{4(aq)}$, so the concentration of $K^+{}_{(aq)}$ is **0.0800 mol dm^{-3}**.

Example 2

What concentration will result from dissolving 1.463g of NaCl in 250 cm^3 of water?

M_r NaCl = 23.0 + 35.5 = 58.5 g mol^{-1} [Na=23.0, Cl=35.5]

Moles of NaCl = 1.463 / 58.5 = 0.0250 moles.

This amount of sodium chloride is dissolved in 250 cm^3 of water.

Concentrations are in 1000 cm^3 so the concentration is 4 times higher (or 1000 / 250 = 4)

Concentration of NaCl = 0.0250 x 4 = **0.100 mol dm^{-3}**.

Example 3

What concentration would result from dissolving 4.51g of $Na_2CO_3.10H_2O$ in 85.0 cm^3 of water?

[Na = 23.0, C = 12.0, O = 16.0, H_2O = 18]

M_r = 2 x 23.0 + 12.0 + 3 x 16.0 + 10 x 18 = 286.0 g mol^{-1}.

Moles of sodium carbonate decahydrate = 4.51 / 286 = 0.01577 moles in 85.0 cm^3 of water.

Concentration of Na_2CO_3 = 0.01577 x 1000 / 85.0 = **0.186 mol dm^{-3}**.

Example 4

What volume of a 0.0750 mol dm^{-3} solution of $HNO_{3(aq)}$ contains 2.45 x 10^{-3} moles?

Volume = moles x 1000 / concentration = 2.45 x 10^{-3} x 1000 / 0.0750 = **32.7 cm^3 of solution**.

[Note: in this question the substance itself doesn't matter, so the M_r of HNO_3 is not used.]

Questions to Test Your Knowledge

Question 1

How many moles of the named substance are there in the following solutions?

a) 250.0 cm^3 of 0.500 mol dm^{-3} sulfuric acid

b) 50 cm^3 of 0.0025 mol dm^{-3} silver nitrate

c) 7.35 cm^3 of 0.15 mol dm^{-3} copper chloride

d) 0.0550 cm^3 of 0.00500 mol dm^{-3} potassium bromide

Question 2

What concentration will result from dissolving the following?

a) 9.53 g of $MgCl_2$ in 1000 cm^3 of water

b) 73.0 g of HCl in 1000 cm^3 of water

c) 0.152 g of $AuCl_3$ in 250 cm^3 of water

d) 1.403 g of KOH in 100.0 cm^3 of water

e) 0.00675 g of $AgNO_3$ in 45 cm^3 of water

f) 1.403 g of $UO_2(NO_3)_2.6H_2O$ in 382 cm^3 of water

Question 3

What volume of the solutions below contain 0.0500 moles?

a) 0.0500 mol dm^{-3} solution of $HNO_{3(aq)}$

b) 2.00 mol dm^{-3} solution of $KNO_{3(aq)}$

c) 0.00650 mol dm^{-3} solution of $HBr_{(aq)}$

d) 4.55 mol dm^{-3} solution of $H_2SO_{4(aq)}$

e) 6.55 x 10^{-4} mol dm^{-3} solution of $Ca(OH)_{2(aq)}$

Question 4

What will be the concentration of the named ion in these solutions?

a) Chloride ion in 1.00 mol dm^{-3} solution of $HCl_{(aq)}$

b) Hydrogen ion in 0.050 mol dm^{-3} solution of $HNO_{3(aq)}$

c) Bromide ion in 0.50 mol dm^{-3} solution of $CuBr_{2(aq)}$

d) Potassium ion in 0.160 mol dm^{-3} solution of $K_3PO_{4(aq)}$

e) Ammonium ion in 0.020 mol dm^{-3} solution of $(NH_4)_2SO_{4(aq)}$

f) Sulfate ion in 0.750 mol dm^{-3} solution of $(NH_4)_2SO_4.FeSO_{4(aq)}$

Question 5

Potassium vanadium alum is a dark violet coloured crystalline ionic substance with the formula $KV(SO_4)_2.12H_2O$.

It is produced by allowing crystals to form from an equimolar solution of potassium sulfate and vanadium(III) sulfate.

$$V_2(SO_4)_{3(aq)} \quad + \quad K_2SO_{4(aq)} \quad \rightarrow \quad 2\ KV(SO_4)_{2(aq)}$$

In the preparation of this substance, 6.66 g of $V_2(SO_4)_3$ was dissolved in 100 cm^3 of water. This was mixed with the required volume of 0.250 mol dm^{-3} $K_2SO_{4(aq)}$.

a) How many moles of vanadium sulfate are there in 6.66 g?

b) How many moles of $K_2SO_{4(aq)}$ will be required?

c) What volume of $K_2SO_{4(aq)}$ will need to be added to make the vanadium alum solution?

d) What mass of vanadium alum crystals will be made?

Question 6

Mohr's salt is a double salt made by crystallising equimolar amounts of $FeSO_4$ and $(NH_4)_2SO_4$ from aqueous solution. Mohr's salt has the formula $(NH_4)_2SO_4.FeSO_4.6H_2O$.

Iron(II) sulfate is made by reacting iron powder with dilute sulfuric acid in the reaction below.

$$H_2SO_{4(aq)} \ + \ Fe_{(s)} \quad \rightarrow \quad FeSO_{4(aq)} \ + \ H_{2(g)}$$

a) How many moles of iron sulfate would be made from the complete reacting of 1.395 g of iron filings?

b) How many cm^3 of hydrogen gas would be released in this process?

Ammonium sulfate is made by neutralising ammonia solution with sulfuric acid according to the reaction below.

$$H_2SO_{4(aq)} \ + \ 2\ NH_{3(aq)} \quad \rightarrow \quad (NH_4)_2SO_{4(aq)}$$

c) What volume of 0.500 mol dm^{-3} $NH_{3(aq)}$ contains 0.100 moles of ammonia?

d) What volume of 1.00 mol dm^{-3} $H_2SO_{4(aq)}$ would you need to exactly neutralise this volume of ammonia solution from part c)?

e) Explain why the $(NH_4)_2SO_{4(aq)}$ solution is in excess?

f) What is the maximum mass of $(NH_4)_2SO_4.FeSO_4.6H_2O_{(s)}$ that could be made from the two solutions prepared as above?

Question 7

Lead iodide is an insoluble yellow solid with the formula PbI_2.

It can be produced by reacting potassium iodide solution with lead ions in solution.

$$Pb^{2+}_{(aq)} \quad + \quad 2\,KI_{(aq)} \quad \rightarrow \quad PbI_{2(s)} \quad + \quad 2\,K^{+}_{(aq)}$$

What mass of $PbI_{2(s)}$ would be precipitated from 45.0 cm^3 of 0.225 mol dm^{-3} $KI_{(aq)}$ by an excess of lead ions?

Part 2 - Acid–Alkali Titration Calculations

Titrations are used to determine a particular unknown quantity by reacting chemicals together until one runs out. This is termed the endpoint. While some titrations show this endpoint by one of the coloured chemicals running out, many need an indicator added that changes colour to highlight the endpoint.

Equation to Learn:

$$Moles = concentration\ x\ \frac{volume\ in\ cm^3}{1000}$$

$$Moles = \frac{C\ x\ V}{1000}$$

$$Moles = C\ x\ V\ (with\ V\ in\ dm^3)$$

- A titre is the volume delivered from the burette
- Pipettes deliver the other fixed volume to the conical flask
- Most indicators have two colours, the endpoint is the halfway colour
- Phenolphthalein changes between pink and colourless and gives the sharpest endpoints

Expected prior knowledge – you should be able to:

- Construct correct formulae
- Construct balanced chemical equations
- Rearrange mathematical equations
- Select appropriate numbers of significant figures for an answer from the data within the question
- Calculate moles from amounts of mass using A_r and M_r
- Calculate moles from concentrations and volume as above
- Calculate percentage purity using the equation:

$$Percentage\ purity = \frac{mass\ of\ identified\ substance}{total\ collected\ mass}\ x\ 100$$

Example 1

In a titration to determine the concentration of hydrochloric acid, a 20.00 cm^3 portion of the $HCl_{(aq)}$ was titrated with 0.150 mol dm^{-3} $NaOH_{(aq)}$. The mean titre was 15.80 cm^3.

Calculate the concentration of the $HCl_{(aq)}$.

There are 4 steps to calculating the missing quantity. REMEMBER these!

These are abbreviated to **EMMA**:

- Equation
- Moles you know
- Moles you don't know
- Answer

E: Write the balanced chemical equation out.

M: Calculate the moles for the chemical that you know both volume and concentration for.

M: Use the reaction ratio to determine the number of moles of the other chemical.

A: Calculate the unknown quantity using the equation for moles, volume and concentration.

$$HCl_{(aq)} + NaOH_{(aq)} \rightarrow NaCl_{(aq)} + H_2O_{(l)}$$

E: See above

M: You know volume and concentration of the NaOH

Moles = 0.150 x 15.80 / 1000 = 0.00237 moles of NaOH

M: Reaction ratio HCl : NaOH = 1 : 1 so moles of HCl = 0.00237 moles

A: Moles = C x V / 1000 so C = Moles x 1000 / V

Concentration of HCl = 0.00237 x 1000 / 20.00 = 0.1185
= **0.119 mol dm^{-3} to 3 s.f.**

Example 2

In a titration to determine the concentration of sodium hydroxide, a 10.00 cm^3 portion of the $NaOH_{(aq)}$ was titrated with 0.500 mol dm^{-3} $H_2SO_{4(aq)}$. The mean titre was 12.55 cm^3.

Calculate the concentration of the $NaOH_{(aq)}$.

$$H_2SO_{4(aq)} + 2NaOH_{(aq)} \rightarrow Na_2SO_{4(aq)} + 2H_2O_{(l)}$$

E: See above

M: You know the volume and concentration of the H_2SO_4.

Moles = 0.50 x 12.55 / 1000 = 0.006275 moles of NaOH

M: Reaction ratio H_2SO_4 : NaOH = 1 : 2 so moles of NaOH = 2 x 0. 006275 = 0.01255 moles

A: Moles = C x V / 1000 so C = Moles x 1000 / V

Concentration of NaOH = 0.01255 x 1000 / 10.00 = 1.255 = **1.26 mol dm^{-3} to 3 s.f.**

Example 3

What will the concentration of un-neutralised sodium hydroxide be if 50 cm^3 of 0.100 mol dm^{-3} $HCl_{(aq)}$ is reacted with 75 cm^3 of 0.250 mol dm^{-3} $NaOH_{(aq)}$?

Chemical Equation:

$$HCl_{(aq)} + NaOH_{(aq)} \rightarrow NaCl_{(aq)} + H_2O_{(l)}$$

Moles of HCl = 0.100 x 50 / 1000 = 0.00500 moles

Moles NaOH = 0.250 x 75 / 1000 = 0.01875 moles

Reaction ratio is 1 : 1 so moles of NaOH neutralised = 0.00500

Unreacted moles of NaOH = 0.01875 – 0.00500 = 0.01375 moles in

Volume = 50 + 75 cm^3 = 125 cm^3.

Concentration of NaOH = 0.01375 x 1000 / 125 = **0.110 mol dm^{-3}**

Example 4 – Back Titration

A **back titration** is where you add an excess of either an acid or an alkali. Part of this reacts away. You then titrate to find the amount remaining and work back to deduce the original amount.

An 8.450 g sample of rock containing magnesium carbonate ($MgCO_3$) was crushed and then reacted with 100 cm^3 of 2.00 mol dm^{-3} $HCl_{(aq)}$ (an excess).

The resulting solution was filtered into a 250.0 cm^3 volumetric flask and the residue washed with distilled water into the flask. The solution was then made up to exactly 250.0 cm^3 with distilled water.

25.00 cm^3 portions of this solution were titrated with 0.200 mol dm^{-3} $NaOH_{(aq)}$.

The mean titre was 19.93 cm^3.

[Mg = 24.3, C = 12.0, O = 16.0]

Determine the percentage by mass of $MgCO_3$ in the rock sample.

You may assume the parts of the rock that are not $MgCO_3$ do not react with acid.

Equations:

Dissolving $MgCO_{3(s)} + 2\,HCl_{(aq)} \rightarrow MgCl_{2(aq)} + CO_{2(g)} + H_2O_{(l)}$

Titration $NaOH_{(aq)} + HCl_{(aq)} \rightarrow NaCl_{(aq)} + H_2O_{(l)}$

Moles of NaOH = 0.2 x 19.93 / 1000 = 0.003986 moles

Moles of HCl titrated = 0.003986 (1:1 reaction in titration)

Moles of HCl in whole 250 cm^3 volumetric flask = 0.003986 x 10 = 0.03986 moles

Moles of HCl added to rock sample = 2.00 x 100 / 1000 = 0.200 moles

Moles of HCl reacted with the $MgCO_3$ in the rock sample = 0.200 – 0.03986
= 0.16014 moles

Moles of $MgCO_3$ = 0.5 x 0.16014 = 0.08007 moles

$M_r[MgCO_3]$ = 24.3 + 12 + 16x3 = 84.3

Mass of $MgCO_3$ in rock sample = 0.08007 x 84.3 = 6.7499 g = 6.750 g

Percentage by mass = 100 x 6.750 / 8.450 = 79.882% = **79.9% to 3 s.f.**

Example 5 – Determining the Unknown Element.

These questions are quite popular with examination boards at A-level.

In these you have to determine the Atomic Mass of an element based upon the titration of a sample containing it.

A 0.650 g sample of a pure Group 1 hydrogen carbonate was dissolved in distilled water and the whole sample titrated with 0.100 mol dm^{-3} $HCl_{(aq)}$. The volume of hydrochloric acid required was 33.52 cm^3.

By calculation, determine the identity of the Group 1 metal.

Equation:

$$MHCO_3 + HCl \rightarrow MCl + CO_2 + H_2O$$

Moles of HCl = 0.100 x 33.52 / 1000 = 0.003352 moles.

Formula is $MHCO_3$ so M_r of HCO_3^- ion is 1 + 12 + 3x16 = 61

Mass from hydrogen carbonate ions = 61 x 0.003352 = 0.204472 g

Mass from Group 1 metal = 0.650 – 0.204472 = 0.445528 g

Moles of Group 1 metal = 0.003352 moles

A_r of Group 1 metal = 0.445528 / 0.003352 = 132.9

The metallic element with an A_r of 132.9 is Caesium.

Note: *When the equation and the formula are not 1:1, then you need to allow for this in the calculation.*

Questions to Test Your Knowledge

Question 8

In a titration to determine the concentration of hydrochloric acid, a 25.00 cm^3 portion of the $HCl_{(aq)}$ was titrated with 0.0250 mol dm^{-3} $NaOH_{(aq)}$.

The mean titre was 18.45 cm^3.

Calculate the concentration of the $HCl_{(aq)}$.

Question 9

In a titration to determine the concentration of potassium hydroxide, a 25.00 cm^3 portion of 0.500 mol dm^{-3} $H_3PO_{4(aq)}$ was titrated with $KOH_{(aq)}$.

The mean titre was 9.50 cm^3.

Calculate the concentration of the $KOH_{(aq)}$.

Question 10

In a titration to determine the concentration of a sodium carbonate solution, a 20.00 cm^3 portion of 0.00500 mol dm^{-3} $H_2SO_{4(aq)}$ was titrated with $Na_2CO_{3(aq)}$ of unknown concentration.

The mean titre was 18.25 cm^3.

Calculate the concentration of the $Na_2CO_{3(aq)}$.

Question 11

In an experiment to determine the mass of calcium hydroxide dissolved to make lime water (saturated $Ca(OH)_{2(aq)}$), 10.00 cm^3 portions were titrated with 0.0500 mol dm^{-3} $HCl_{(aq)}$. If the mean titre was 9.36 cm^3, calculate the solubility of calcium hydroxide in water in:

a) mol dm^{-3}

b) g dm^{-3}

Question 12

A sample of $K_2CO_{3(s)}$ was dissolved in exactly 100.0 cm^3 of water.

10.00 cm^3 portions of this solution were titrated with 0.0500 mol dm^{-3} $HCl_{(aq)}$.

If the mean titre was 13.50 cm^3, what was the mass of the potassium carbonate in the sample?

Question 13

Calculate the missing values from the titrations in the table below.

Expt	Pipette contained	Pipetted concentration / mol dm^{-3}	Pipette volume / cm^3	Burette contained	Burette concentration / mol dm^{-3}	Mean titre / cm^3
A	$HCl_{(aq)}$	0.0500	20.00	$KOH_{(aq)}$		13.65
B	$Ca(OH)_{2(aq)}$	0.00575	25.00	$HNO_{3(aq)}$		9.85
C	$LiOH_{(aq)}$	0.755	10.00	$H_2SO_{4(aq)}$		44.45
D	$Na_2CO_{3(aq)}$	0.250	25.00	$HCl_{(aq)}$	0.625	
E	$NaHCO_{3(aq)}$	0.750		$H_2SO_{4(aq)}$	0.475	15.79

Question 14

The mineral celestine contains strontium sulfate. It can be converted into strontium oxide by the reaction sequence below.

Reaction 1: $SrSO_{4(aq)} + 4\,C_{(s)} \rightarrow SrS_{(s)} + 4\,CO_{(g)}$

Reaction 2: $SrS_{(s)} + 2\,NaOH_{(aq)} \rightarrow Sr(OH)_{2(s)} + Na_2S_{(aq)}$

Reaction 3: $Sr(OH)_{2(s)} \rightarrow SrO_{(s)} + H_2O_{(l)}$

a) What potential hazard is there in the treatment of strontium sulfate in reaction 1?

b) Suggest how reaction 3 would be carried out.

c) 12.5 g of strontium oxide was obtained from 46.2 g of strontium sulfate in this reaction sequence.
Calculate the yield of the three-reaction sequence.

A 0.256 g sample of impure SrO was dissolved in exactly 100.0 cm^3 of water. The entire solution was titrated with 0.500 mol dm^{-3} $HCl_{(aq)}$.

The titre required was 9.60 cm^3.

d) Calculate the purity of the SrO in the 0.256 g sample as a percentage.

Question 15

A 0.267 g sample of a pure Group 1 hydrogen carbonate ($MHCO_3$) was dissolved in distilled water and the whole sample titrated with 0.150 mol dm^{-3} $HCl_{(aq)}$.

The volume of hydrochloric acid required was 21.19 cm^3.

Determine the identity of the metal in the Group 1 hydrogen carbonate.

Question 16

A 0.350 g sample of a pure Group 1 carbonate (M_2CO_3) was dissolved in distilled water and the whole sample titrated with 0.200 mol dm^{-3} $HCl_{(aq)}$.

The volume of hydrochloric acid required was 25.33 cm^3.

Determine the identity of the metal in the Group 1 carbonate.

Question 17

Rhubarb leaves and stalk contain ethanedioic acid.

Ethanedioic acid has the structure $(CO_2H)_2$ and is a diprotic acid.

The amount of ethanedioic acid in rhubarb leaves was determined by the following method.

1. 50.0 g of raw rhubarb leaves were chopped in a blender with 25.00 cm^3 of 0.0500 mol dm^{-3} $H_3PO_{4(aq)}$ and approximately 25 cm^3 of distilled, de-ionised water.
2. The resulting mixture was then filtered and the residue washed with a further 25 cm^3 of distilled, de-ionised water.
3. The solution was then made up to exactly 100.0 cm^3 with more distilled, de-ionised water.
4. The extract solution was left for 24 hours to allow any fine sediment to fall to the bottom.
5. 25.00 cm^3 portions of this extract solution was then titrated with 0.250 mol dm^{-3} $KOH_{(aq)}$ with a mean titre of 16.70 cm^3.

Calculate the mass of ethanedioic acid per 100 g of rhubarb leaves.

Question 18

A 0.166 g sample of a pure Group 2 carbonate (MCO_3) was found to react exactly with 15.75 cm^3 of 0.250 mol dm^{-3} $HNO_{3(aq)}$.

Determine the identity of the metal in the Group 2 carbonate.

Question 19

A 0.850 g sample of a pure Group 1 carbonate (M_2CO_3) was dissolved in about 50 cm^3 of distilled water in a beaker.

This solution was transferred to a volumetric flask, the beaker washings added to the volumetric flask and the volume was then made up to exactly 100.00 cm^3. 10.00 cm^3 portions were titrated with 0.0500 mol dm^{-3} $H_2SO_{4(aq)}$.

The volume of sulfuric acid required was 23.04 cm^3.

Determine the identity of the metal in the Group 1 carbonate.

Question 20

Phosphorous pentachloride ($PCl_{5(s)}$) reacts with water to make $HCl_{(aq)}$.

A 0.875 g sample of impure PCl_5 was added to water.

PCl_5 reacts with water to produce a solution of hydrochloric acid.

$$PCl_{5(s)} + H_2O_{(l)} \rightarrow POCl_{3(l)} + 2\ HCl_{(aq)}$$

The $POCl_3$ produced reacts with water to release further HCl.

$$POCl_{3(l)} + 3\ H_2O_{(l)} \rightarrow H_3PO_{4(aq)} + 3HCl_{(aq)}$$

The resulting solution was diluted to a volume of exactly 250.0 cm^3.

20.00 cm^3 portions of this solution were titrated with 0.0500 mol dm^{-3} $NaOH_{(aq)}$ with a mean titre of 27.66 cm^3.

a) Calculate the mass of phosphorous pentachloride in the 0.515 g original sample and use this to calculate its purity.

b) What assumption about the impure PCl_5 have you made?

Examination Style Questions

Question 1

For the neutralisation shown below:

$$HCl_{(aq)} + NaOH_{(aq)} \rightarrow NaCl_{(aq)} + H_2O_{(l)}$$

A 25.00 cm^3 sample of 0.550 mol dm^{-3} $HCl_{(aq)}$ was titrated with a solution of $NaOH_{(aq)}$ of unknown concentration. The mean titre was 18.55 cm^3.

a) Calculate the concentration of the sodium hydroxide. (4)

Concentration to 3 s.f = ...mol dm^{-3}

b) State the expected colour change when using phenolphthalein as the indicator in this titration. (1)

....

Total = 5 marks

Question 2

For the neutralisation shown below:

$$H_2SO_{4(aq)} + 2\ KOH_{(aq)} \rightarrow K_2SO_{4(aq)} + 2\ H_2O_{(l)}$$

A 20.00 cm^3 sample of 0.850 mol dm^{-3} $H_2SO_{4(aq)}$ was titrated with a solution of $KOH_{(aq)}$ of unknown concentration. The mean titre was 37.25 cm^3.

a) Calculate the concentration of the potassium hydroxide. (4)

Concentration = ...mol dm^{-3}

b) State the expected colour change when using methyl orange as the indicator in this titration. (1)

....

Total = 5 marks

Question 3

For the neutralisation shown below:

$$2\,HCl_{(aq)} \quad + \quad Ca(OH)_{2(aq)} \quad \rightarrow \quad CaCl_{2(aq)} \quad + \quad 2\,H_2O_{(l)}$$

A 15.00 cm^3 sample of 0.0220 mol dm^{-3} $Ca(OH)_{2(aq)}$ was titrated with a solution of $HCl_{(aq)}$ of unknown concentration. The mean titre was 25.05 cm^3.

a) Calculate the concentration of the hydrochloric acid. (4)

Concentration = ... mol dm^{-3}

b) Suggest a reason why you could not use twice the concentration of calcium hydroxide? (1)

....

c) Explain why methyl orange is a better choice of indicator than phenolphthalein. (2)

....

....

....

....

Total = 6 marks

Question 4

A 10.00 cm^3 sample of 1.250 mol dm^{-3} $H_3PO_{4(aq)}$ was titrated with a solution of $KOH_{(aq)}$ of unknown concentration. The mean titre was 43.25 cm^3.

a) Write the equation for the neutralisation. Include state symbols in your answer. (2)

....

b) Calculate the number of moles of phosphoric acid. (1)

c) Deduce the number of moles of potassium hydroxide. (1)

d) Calculate the concentration of the hydrochloric acid. (1)

Concentration = ...mol dm^{-3}

Total = 5 marks

Question 5

A 25.00 cm^3 sample of 0.00500 mol dm^{-3} $H_2SO_{4(aq)}$ was titrated with a solution of $Ca(OH)_{2(aq)}$ of unknown concentration. The mean titre was 22.95 cm^3.

a) Calculate the concentration of the calcium hydroxide solution. (4)

Concentration = ...mol dm^{-3}

Total = 4 marks

Question 6

A 25.00 cm^3 sample of 1.50 mol dm^{-3} $HCl_{(aq)}$ was titrated with 0.9000 mol dm^{-3} $NaOH_{(aq)}$.

a) Calculate the expected titre of sodium hydroxide solution. (3)

Titre = ... cm^3

b) State the expected colour change when using phenolphthalein as the indicator in this titration. (1)

....

Total = 4 marks

Question 7

A 15.00 cm^3 sample of 1.25 mol dm^{-3} sulfuric acid was titrated with 2.25 mol dm^{-3} $KOH_{(aq)}$.

a) Calculate the expected titre of potassium hydroxide solution. (3)

Titre = ... cm^3

b) Explain why both methyl orange and phenolphthalein could be used in this titration. (2)

....

....

....

....

Total = 5 marks

Question 8

A 40.00 cm^3 sample of 0.00375 mol dm^{-3} $Ca(OH)_{2(aq)}$ was titrated with 0.0850 mol dm^{-3} $HCl_{(aq)}$.

a) Calculate the expected titre of hydrochloric acid solution. (3)

Titre = ... cm^3

Total = 3 marks

Question 9

A sample of 0.710 mol dm^{-3} $H_3PO_{4(aq)}$ containing 0.01420 moles of phosphoric acid was pipetted into a conical flask and titrated with 0.0950 mol dm^{-3} $KOH_{(aq)}$.

a) Calculate the expected titre of potassium hydroxide solution. (3)

Titre = ... cm^3

b) Calculate the volume of the pipette used for the titration. (2)

Pipette size = ... cm^3

Total = 5 marks

Question 10

An entire 100.0 cm^3 volumetric flask of 0.000685 mol dm^{-3} $Ca(OH)_{2(aq)}$ was titrated with 0.0250 mol dm^{-3} $H_2SO_{4(aq)}$.

a) Calculate the expected titre of sulfuric acid solution. (3)

Concentration to = ... mol dm^{-3}

b) What would the titre be if the sulfuric acid used had a concentration of 0.00500 mol dm^{-3}? (1)

Titre = ...cm^3

c) Explain why a sulfuric acid concentration of 0.00500 mol dm^{-3} introduces less uncertainty into the volume of the titre. (2)

....

....

....

....

Total = 6 marks

Question 11

To find the concentration of a solution of sodium hydroxide, 20.00 cm^3 portions of the solution were titrated with a known concentration of hydrochloric acid.

a) Write out the full equation for the reaction of sodium hydroxide with hydrochloric acid in aqueous solution including state symbols. (2)

....

b) What is the reaction ratio between NaOH and HCl? (1)

Ratio: ....

c) Explain the term **concordant titres**. (1)

....

....

Titre	Volume of 0.0500 mol dm^{-3} $HCl_{(aq)}$ / cm^3
Range finder	23.0
1	22.85
2	23.15
3	22.85
4	22.95

d) Identify which of the results in the table above should be used in the calculation. (1)

Results used: ....

e) Calculate the mean titre. (1)

Mean titre: .... cm^3

f) Use your value for the mean titre to calculate a value for the concentration of the $NaOH_{(aq)}$ solution. (3)

Concentration of NaOH(aq) : .... mol dm^{-3}

Total = 9 marks

Question 12

500.0 cm^3 of 0.250 mol dm^{-3} solution of sodium hydrogen carbonate needs to be made.

a) Describe the quantities and equipment required as well as the process to accurately make up the solution. (4)

....

....

....

....

....

....

....

....

....

....

....

....

....

....

....

....

b) Write the full equation for the reaction of sodium hydrogen carbonate with nitric acid. (2)

....

c) Convert your equation in part b) to an ionic equation. Include the correct state symbols. (2)

....

d) State the name of a suitable indicator and the expected colour change when sodium hydrogen carbonate is titrated with nitric acid. Explain the role of your chosen indicator in the titration. (3)

Indicator choice: ....

Expected colour change: ....

Indicator role: ....

....

....

e) What other observation would you expect to make as the titration was carried out? (1)

....

Total = 12 marks

Question 13

A titration was carried out between 0.0150 mol dm^{-3} phosphoric acid ($H_3PO_{4(aq)}$) and a saturated solution of calcium hydroxide to determine the solubility of calcium hydroxide in water.

The phosphoric acid solution was made up in a volumetric flask using a mass of pure phosphoric acid measured on a 2 d.p. balance.

a) Complete the equation for the neutralisation that shows phosphoric acid and calcium hydroxide react in a 2:3 ratio.
State symbols are **NOT** required. (2)

……H_3PO_4 + ……$Ca(OH)_2$ $\rightarrow$ ....

The errors in using the equipment are listed below.

Balance...	± 0.005 g	*Read twice to find mass*
25 cm^3 pipette	± 0.06 cm^3	*Read once in using*
10 cm^3 pipette..............................	± 0.03 cm^3	*Read once in using*
Class A Burette............................	± 0.02 cm^3	*Read twice in a titration*
Class B Burette............................	± 0.05 cm^3	*Read twice in a titration*
250 cm^3 volumetric flask................	± 0.3 cm^3	*Read once in filling*
100 cm^3 volumetric flask.................	± 0.2 cm^3	*Read once in filling*

Two students carried out the experiment using variations of the equipment.

Student 1	***Student 2***
250 cm^3 volumetric flask 25 cm^3 pipette Class A Burette Balance	100 cm^3 volumetric flask 10 cm^3 pipette Class B Burette Balance
Mass of phosphoric acid used = 0.37 g Titre = 17.0 cm^3	Mass of phosphoric acid used = 0.15 g Titre = 6.8 cm^3

b) Complete the errors table for both students and calculate a total percentage error for the titration. (5)

Student 1	**Percentage error**	**Student 2**	**Percentage error**
Balance		Balance	
Volumetric flask		Volumetric flask	
Pipette		Pipette	
Burette		Burette	
Total % Uncertainty		**Total % Uncertainty**	**± 8.64 %**

c) State which method has the lowest total percentage error and identify the main cause of uncertainty within both the student's methods. (2)

Student with the lowest percentage error: ....

Main cause of uncertainty: ....

....

....

....

The value for the solubility of calcium hydroxide is 1.13 g dm^{-3}.

d) Complete the table of final results for each student. (3)

	Student 1	**Student 2**
Total percentage uncertainty		**± 8.64 %**
Solubility of $Ca(OH)_2$	1.13 g dm^{-3}	1.13 g dm^{-3}
Uncertainty in g dm^{-3}	± g dm^{-3}	± g dm^{-3}
Range of values	1.09 to 1.17 g dm^{-3}	to g dm^{-3}

The saturated solution of calcium hydroxide was found to have a sediment of undissolved $Ca(OH)_{2(s)}$ at the bottom.

e) Describe the effect of including some of this sediment on the titre of phosphoric acid and the calculated solubility of calcium hydroxide from the experiment. (2)

Effect on titre in titration: ....

....

....

....

Effect on calculated value of solubility: ....

....

....

....

Total = 14 marks

Question 14

Tartaric acid is a natural dibasic acid obtained from grapes, bananas and tamarind fruits.

$M_{r[\text{Tartaric acid}]} = 150$.

The amount of tartaric acid in tamarind pulp was determined using the following method:

1. 175 g of tamarind pulp was liquidised with 15.0 cm^3 of 0.650 mol dm^{-3} phosphoric acid to extract the tartaric acid.
2. The resulting mixture was filtered into a 250.0 cm^3 volumetric flask.
3. The liquidiser was rinsed out with distilled water and added to the filter funnel.
4. The pulp was then washed with more distilled water into the volumetric flask.
5. The volumetric flask was then made up to the calibration mark with distilled water.
6. 25.00 cm^3 of this solution was titrated with 0.350 mol dm^{-3} potassium hydroxide solution with a mean titre of 18.87 cm^3.

a) Calculate the mass of tartaric acid in 175 g of tamarind pulp. (8)

b) Express this as a percentage by mass. (1)

Total = 9 marks

Question 15

2.33 g of calcium hydroxide was dissolved in approximately 100 cm^3 of distilled water in a 250 ml beaker. *(The volume of water is therefore not known accurately.)*

A titration of 25.00 cm^3 of this solution of calcium hydroxide of unknown strength was carried out with 0.0250 mol dm^{-3} dilute hydrochloric acid.

The following results were obtained.

Titration	**Rangefinder**	**Titre 1**	**Titre 2**	**Titre 3**
Result / cm^3	13	12.60	12.95	12.70

a) Which **TWO** results should be selected as concordant titres? (1)

....

b) Write the full balanced equation for the reaction of calcium hydroxide with hydrochloric acid including state symbols. (2)

....

c) Calculate the concentration of the calcium hydroxide solution using a mean value from your selected concordant titres.
Give your answer to 3 significant figures. (3)

Concentration of calcium hydroxide solution = ... mol dm^{-3}.

d) Name the type of error when someone who is short looks up to the level of liquid in the burette when setting the value to 0.00 cm^3. (1)

....

e) The balance that was used in making up the solutions has a stated error of ± 0.005 g. Given this error, what is the possible range that a reading of 2.33 g would have? (1)

....

The uncertainties with the pieces of glassware and other equipment used in the experiment have the values shown in the table below.

Equipment	Capacity	Uncertainty	Percentage uncertainty
Burette	50.0 cm^3	± 0.05 cm^3	
Pipette	25.0 cm^3	± 0.06 cm^3	± 0.24 %
Volumetric flask	5000 cm^3	± 5 cm^3	± 0.10 %
Balance	600.00 g	± 0.005 g	± 0.43 %

f) Calculate the percentage uncertainty for using the burette in a titration using the mean value of your chosen concordant titres in part a). Add it to the table above. (1)

g) Calculate a total percentage uncertainty and then use this to state the concentration of the calcium hydroxide solution with its uncertainty in mol dm^{-3} to 5 decimal places. (2)

Concentration of calcium hydroxide solution = ...mol dm^{-3}.

Total = 11 marks

Question 16

Lava from the Ol Doinyo volcano in Tanzania is unusual in that it erupts a lava rich in carbonates due to the low temperature of the emerging lava.

The lava contains a mixture of nyerereite $Na_2Ca(CO_3)_2$ with other minerals.

45.0 g of the lava was crushed and reacted with 200.0 cm^3 of 1.00 mol dm^{-3} hydrochloric acid (an excess of $HCl_{(aq)}$).

$$Na_2Ca(CO_3)_{2(s)} + 4\,HCl_{(aq)} \rightarrow 2\,NaCl_{(aq)} + CaCl_{2(aq)} + 2\,CO_{2(g)} + 2\,H_2O_{(l)}$$

The solution containing the unreacted acid was then filtered into a volumetric flask and made up to exactly 250.0 cm^3.

20.00 cm^3 portions of this solution were titrated with 0.0100 mol dm^{-3} NaOH.

The mean titre was 28.02 cm^3.

a) Calculate the percentage by mass of nyerereite in the sample of lava. Give your answer to an appropriate number of significant figures. (8)

(You can assume the other minerals do not react with the hydrochloric acid.)

Percentage mass $Na_2Ca(CO_3)_2$ = ... %

b) State the effect on the uncertainty of your answer if the sodium hydroxide used had a concentration of 0.0200 mol dm^{-3}. (2)

....

....

Total = 10 marks

Question 17

Indigestion tablets frequently contain calcium carbonate and magnesium hydroxide in their formulation.

a) Assuming stomach acid to be hydrochloric acid (HCl), complete the second equation for $Mg(OH)_2$ neutralising stomach acid. State symbols are **NOT** required. (2)

Reaction 1:

$$CaCO_3 + 2\,HCl \rightarrow CaCl_2 + H_2O$$

Reaction 2:

$$Mg(OH)_2 + HCl \rightarrow \text{..}$$

A formulation called **EVXS** contains 0.500 g of $CaCO_3$ and 0.580 g of $Mg(OH)_2$ as anhydrous solids.

b) One **EVXS** tablet neutralises exactly 60.75 cm^3 of stomach acid.

Work out the moles of acid that each of the **two** component chemicals will neutralise separately. Use these answers to deduce the concentration of the stomach acid. Give your answer to 3 significant figures. (6)

Concentration of stomach acid .... mol dm^{-3}

Total = 8 marks

Question 18

Sulfamic acid is an acidic molecule with the formula H_3NSO_3.

Sulfamic acid is a monoprotic acid.

Sulfamic acid is used as a descaling agent to remove precipitated $CaCO_{3(s)}$.

One **EVDS** descaling tablet for an espresso coffee machine has a mass of 5.00 g and contains sulfamic acid along with additional neutral cleaning agents.

a) Write the balanced chemical equation for the reaction of sulfamic acid with calcium carbonate. (2)

....

One tablet of **EVDS** was dissolved in exactly 100.00 cm^3 of water.

10.00 cm^3 of this solution was removed and titrated with 0.250 mol dm^{-3} sodium hydroxide.

The mean titre was 11.25 cm^3.

b) Calculate the mass of sulfamic acid in one **EVDS** tablet. (4)

Mass of sulfamic acid in one EVDS tablet = ... g

c) Use your answer to part b) to calculate the mass of calcium carbonate that **one EVDS** tablet will remove using to your chemical equation in part a). (2)

Mass of calcium carbonate removed by **one** EVDS tablet = ... g

Total = 8 marks

Question 19

Benzoic acid (E220) is a food preservative that prevents the growth of moulds and yeasts.

Benzoic acid has the formula $C_7H_6O_2$ and is a solid at room temperature with a low solubility in cold water. Benzoic acid is a monoprotic acid.

A saturated solution of benzoic acid was made up using the following method.

1. Approximately 5 g of benzoic acid was added to 1.00 dm^3 of hot water.
2. The solution was left to cool.
3. The undissolved benzoic acid was removed by filtration.

25.00 cm^3 portions of this solution were added to a conical flask and titrated with 0.0850 mol dm^{-3} $KOH_{(aq)}$. The mean titre was 8.10 cm^3.

a) Calculate the solubility in g dm^{-3}. Give your answer to an appropriate number of significant figures. (4)

Solubility of benzoic acid = ... g dm^{-3}

b) How could you lower the level of uncertainty in the titration volume? Justify your answer. (2)

....

....

c) How would the solubility result be altered if the prepared solution were used at a different temperature to that it was prepared at? (2)

Lower temperature: ....

Higher temperature: ....

d) What would you have to be careful not to do when filling a pipette when the solution had been left at a cooler temperature?
Explain how this would alter the value for the solubility? (2)

....

....

....

....

Total = 10 marks

Question 20

Substance B is a white solid known to be a monoprotic organic acid.

3.785 g of **substance B** were dissolved in 125 cm^3 of ethanol in a volumetric flask. The solution was made up to exactly 250.0 cm^3 with distilled water.

15.00 cm^3 portions of this solution were titrated with 0.0350 mol dm^{-3} $NaOH_{(aq)}$.

The mean titre was 39.56 cm^3.

a) Calculate the M_r of **substance B** using the titration result. (4)

b) State what the M_r of **substance B** would be if it were a diprotic acid. (1)

....

c) Given the results of the combustion analysis below, calculate the empirical formula and state the molecular formula. (5)

Mass burned / g	Mass of carbon dioxide / g	Mass of water / g
1.955	5.245	1.287

Empirical formula: ....

Molecular formula: ....

Total = 10 marks

Question 21

A pure sample of a Group 2 metal oxide was analysed to find the identity of the metal by carrying out a back titration.

3.355 g of the pure metal oxide was reacted with exactly 100.0 cm^3 of 2.00 mol dm^{-3} $HNO_{3(aq)}$.

The resulting solution was diluted to exactly 500.0 cm^3 in a volumetric flask.

25.00 cm^3 portions of this solution were titrated with 0.0400 mol dm^{-3} sodium hydroxide solution.

The mean titre was 41.87 cm^3.

a) Write the balanced general equation for the reaction of a Group 2 oxide with nitric acid using MO to represent the Group 2 metal oxide. State symbols are **NOT** required. (1)

....

b) What is the ratio for the reaction of nitric acid with sodium hydroxide? (1)

....

c) Identify the Group 2 metal in the oxide by calculation. (7)

Identity of the **metal** in MO: ....

d) If 0.100 mol dm^{-3} $NaOH_{(aq)}$ had been used for the titration, calculate the way the uncertainty of the titre would have changed if the same burette (±0.06 cm^3) was used. (3)

....

....

....

....

Total = 12 marks

Question 22

Ethanedioic acid is a white crystalline solid with the formula $H_2C_2O_4.\mathbf{x}H_2O$.

Ethanedioic acid is a weak diprotic acid.

A 1.056 g sample of ethanedioic acid was dissolved in distilled water in a volumetric flask and made up to exactly 200.0 cm^3.

20.00 cm^3 portions of this solution were titrated with 0.0500 mol dm^{-3} sodium hydroxide solution with a mean titre of 33.52 cm^3.

a) Calculate the value of **x** in the formula $H_2C_2O_4.\mathbf{x}H_2O$. (6)

Value of **x** = ...

b) Explain why phenolphthalein would be a suitable indicator for this titration but not methyl orange? (2)

....

....

....

....

The uncertainties in the measurements from the titration procedure are:

Equipment	Measurement	Uncertainty	Percentage uncertainty
Pipette	20.00 cm^3	0.05 cm^3	
Burette	33.52 cm^3	0.06 cm^3	
Volumetric flask	200.0 cm^3	0.25 cm^3	
Balance	1.056 g	0.0005 g	
		Total percentage uncertainty =	**0.83 %**

c) Calculate the **FOUR** missing percentage uncertainties. (4)

d) If the hydrated ethanedioic acid had been gently heated before dissolving, calculate the expected titre assuming that all of the water of crystallisation had been removed before dissolving but the same mass (1.056 g) of this anhydrous ethanedioic acid used. (3)

Expected titre = ... cm^3

Total = 15 marks

Question 23

An ore consisting of the mineral rhodochrosite, $MnCO_3$, was analysed.

The ore contained a low level of $MnCO_3$.

The rest of the mass was carrier rock.

In order to determine the percentage by mass of rhodochrosite in the ore, the following procedure was followed.

1. The ore was crushed to a powder.
2. 25.00 g of the ore was added to 100.0 cm^3 of 0.250 mol dm^{-3} hydrochloric acid and warmed to 80°C to react the $MnCO_3$.
3. Once cooled the mixture was filtered to remove the unreacted carrier rock.
4. The residue was washed with distilled water and the solution made up to exactly 500.0 cm^3 in a volumetric flask.
5. 50.00 cm^3 portions of this solution were titrated with 0.100 mol dm^{-3} sodium hydroxide solution.
6. The mean titre was 13.03 cm^3.

a) Write the balanced equation for the reaction of manganese carbonate with hydrochloric acid. Include state symbols in your answer. (2)

....

b) Calculate the percentage by mass of $MnCO_3$ in the rock sample. Give you answer to an appropriate number of significant figures. (7)

Percentage by mass of $MnCO_3$ in rock sample = ... %

c) What assumption has been made about the carrier rock? (1)

....

....

....

d) Explain the effect on the percentage by mass of not adding the washings in step 4 to the volumetric flask. (2)

....

....

....

....

....

....

....

....

....

Total = 12 marks

Question 24

A sample of a metal hydroxide of formula $M(OH)_2.H_2O$ was titrated to find the identity of the metal using the following method.

1. 1.560 g of $M(OH)_2$ was added to a volumetric flask.
2. The metal hydroxide was reacted with 50.0 cm^3 of 1.00 mol dm^{-3} nitric acid.
3. The resulting solution was made up to 100.0 cm^3 with distilled water.
4. 10.00 cm^3 portions of this solution were titrated with 0.0505 mol dm^{-3} $KOH_{(aq)}$.
5. The mean titre was 31.93 cm^3.

a) By calculation, find the identity of **metal M**. (10)

Identity of **metal M** = ...

b) If 1.545 g of nickel hydroxide, $Ni(OH)_{2(s)}$, was used in the same process with the same chemicals and concentrations, show that the expected titre would be 33.0 cm^3. (5)

c) If using phenolphthalein, give a reason why the end point in this second titration would be difficult to observe. (1)

....

....

Total = 16 marks

Question 25

A sample of iron hydroxide was analysed to determine its composition.

The iron hydroxide has the formula $Fe(OH)_x$ where **x** is either 2 or 3.

The procedure was as follows.

1. A 9.540 g sample of the iron hydroxide was reacted with exactly 150.0 cm^3 of 2.00 mol dm^{-3} nitric acid in a beaker.
2. The resulting solution was transferred to a 250.0 cm^3 volumetric flask. The beaker was washed out into the volumetric flask thoroughly then the volume was made up to exactly 250.0 cm^3 using distilled water.
3. 10.00 cm^3 portions of this solution were titrated with 0.750 mol dm^{-3} $NaOH_{(aq)}$.
4. The mean titre was found to be 4.68 cm^3.

a) Determine the value of **x** in the formula of the iron hydroxide. (7)

Value of **x** in the formula $Fe(OH)_x$ = ...

b) Explain the effect of the iron hydroxide being a mixture of $Fe(OH)_2$ and $Fe(OH)_3$ on your answer to part a). (3)

....

....

....

....

....

....

....

....

c) State a simple step that would reduce the uncertainty in the titre. Explain your answer. (2)

....

....

....

....

Total = 12 marks

Question 26

Iron oxide exists in 3 common forms. FeO, Fe_2O_3 and Fe_3O_4.

a) (i) State the oxidation number of the Fe in each of the oxides. (2)

FeO ....

Fe_2O_3 ....

Fe_3O_4 ....

(ii) Explain the non-integer vales for the oxidation number in Fe_3O_4. (1)

....

....

....

....

....

....

A sample of an iron oxide of unknown formula was analysed using the procedure shown below.

1. A 25.50 g sample of the iron oxide was reacted with 325 cm^3 of 5.00 mol dm^{-3} sulfuric acid.
2. The solution was filtered and filtrate added to a volumetric flask.
3. The glassware used and the residues were rinsed with distilled water into the volumetric flask.
4. The solution in the volumetric flask was made up to 500.0 cm^3 in a volumetric flask with the washing included in the flask.
5. 15.00 cm^3 portions of this solution were titrated with 2.00 mol dm^{-3} $KOH_{(aq)}$.
6. The mean titre was 34.39 cm^3.

b) By calculation, determine which of the three iron oxides was used in the procedure. (8)

Formula of iron oxide = ...

Total = 11 marks

Answers

Do not round answers during calculations, write down answers with too many significant figures. Only round your final answer to the correct number of significant figures as explained below.

In working out answers it is worth remembering the following M_r values.

H_2O = 18.0	O_3 = 48	SO_4 = 96.1 (32.1+3x16)
OH = 17.0	O_4 = 64	CO_3 = 60.0 (12+3x16)
CO_2 = 44.0	Cl_2 = 71.0	NO_3 = 62.0 (14+3x16)

In all answers, the number of significant figures is determined by the least accurate piece of data within the question. This may be the A_r value from a periodic table.

250 is 2 s.f   250.0 is 4 s.f   0.00250 is 3 s.f   0.000025 is 2 s.f.

Questions To Test Your Knowledge – Part 1

<table>
<tr><td>Question 1</td><td colspan="4">a) 0.500 x 250 / 1000 = 0.125 moles to 3 s.f
b) 0.0025 x 50 / 10000 = 1.25 x 10^{-4} = 1.3 x 10^{-4} moles to 2 s.f.
c) 0.15 x 7.35 / 1000 = 0.0011 moles to 2 s.f.
d) 0.00500 x 0.055 / 1000 = 2.70 x 10^{-8} moles to 3 s.f.</td></tr>
<tr><td rowspan="7">Question 2</td><td>Part</td><td>M_r</td><td>Moles</td><td>Concentration</td></tr>
<tr><td>a)</td><td>24.1 + 2x35.5 = 95.3</td><td>9.53 / 95.3 = 0.100</td><td>0.100 x 1000 / 1000
= 0.100 mol dm^{-3}</td></tr>
<tr><td>b)</td><td>1 + 35.5 = 36.5</td><td>73 / 36.5 = 2.00</td><td>2.00 x 1000 / 1000
= 2.00 mol dm^{-3}</td></tr>
<tr><td>c)</td><td>197 + 3x35.5 = 303.5</td><td>0.152 / 303.5 = 0.0005008</td><td>0.0005008 x 1000 / 250
= 0.00200 mol dm^{-3}</td></tr>
<tr><td>d)</td><td>39.1 + 17 = 56.1</td><td>1.403 / 56.1 = 0.025009</td><td>0.025009 x 1000 / 100
= 0.250 mol dm^{-3}</td></tr>
<tr><td>e)</td><td>107.9 + 14 + 48 = 169.9</td><td>0.00675 / 169.9 = 3.973 x 10^{-5}</td><td>3.973 x 10^{-5} x 1000 / 45
= 8.8 x 10^{-4} mol dm^{-3}</td></tr>
<tr><td>f)</td><td>238 + 32 + 2x62 + 6x18
= 502</td><td>1.403 / 502 = 0.002795</td><td>0.002795 x 1000 / 382
= 0.00732 mol dm^{-3}</td></tr>
<tr><td>Question 3</td><td colspan="4">a) 0.0500 x 1000 / 0.0500 = 1000 cm^3
b) 0.0500 x 1000 / 2.00 = 25.0 cm^3
c) 0.0500 x 1000 / 0.00650 =7692.3 cm^3 = 7690 cm^3
d) 0.0500 x 1000 / 4.55 = 10.989 cm^3 = 11.0 cm^3
e) 0.0500 x 1000 / 6.55 x 10^{-4} = 76355.9 cm^3 = 76300 cm^3 or 76.3 dm^3</td></tr>
<tr><td>Question 4</td><td colspan="4">a) 1.00 mol dm^{-3} (Formula contains one Cl^-.)
b) 0.050 mol dm^{-3} (Formula contains one H^+.)
c) 1.0 mol dm^{-3} (Formula contains two Br^-.)
d) 3 x 0.160 = 0.480 mol dm^{-3} (Formula contains three K^+.)
e) 2 x 0.020 = 0.40 mol dm^{-3} (Formula contains two NH_4^+.)
f) 2 x 0.750 = 1.50 mol dm^{-3} (Formula contains two SO_4^{2-}.)</td></tr>
</table>

Question 5	a) M_r = 2 x 50.9 + 3 x 96.1 = 390.1 so 6.6 / 390.1 = **0.017073 moles** b) Equation shows 1 to 1 ratio so **0.017073 moles** c) 0.017073 moles x 1000 / 0.250 = **68.3 cm^3** d) $M_{r[KV(SO_4)_2.12H_2O]}$ = 39.1 + 50.9 + 2x96.1 + 12x18 = 498.2. The balanced equation shows 2 moles of vanadium alum will be made from 1 moles of vanadium sulfate so 2 x 0.017073 = 0.034146 moles is made. The mass made = 498.2 x 0.034146 = 17.01154 g = **17.0 g to 3 s.f.**
Question 6	a) Moles of iron = 1.395 / 55.8 = **0.0250 moles** b) 1 mole of gas at RTP occupies 24000 cm^3 so 0.0250 x 24000 = **600 cm^3** c) 0.100 x 1000 / 0.500 = **200 cm^3** d) Reaction ratio is 1 to 2 so need 0.0500 moles of sulfuric acid 0.0500 x 1000 / 1.00 = **50 cm^3** e) 0.0250 moles of iron(II) sulfate formed. 0.100 moles of ammonia will form 0.0500 moles of ammonium sulfate. Equimolar reaction so a double excess of ammonium sulfate. f) M_r = 2x(14+4) + 96.1 + 55.8 + 96.1 + 6x18 = 392.0 Moles made = 0.0250 so mass = 0.0250 x 392.0 = 9.80 g
Question 7	Moles of KI = 0.0225 x 45 / 1000 = 0.010125 moles Reaction ratio of Pb^{2+} to I^- is 1 : 2 so 0.010125 / 2 = 0.0050625 moles PbI_2 $M_{r[PbI_2]}$ = 207.2 + 2x126.9 = 461.0 Mass of PbI_2 formed = 461.0 x 0.0050625 = 2.3338 g = **2.33 g to 3 s.f.**

Questions To Test Your Knowledge – Part 2

Question 8	$HCl + NaOH \rightarrow NaCl + H_2O$ Moles of NaOH = 0.0250 x 18.45 / 1000 = 4.6125×10^{-4} moles Ratio 1:1 Moles HCl = 4.6125×10^{-4} moles Concentration HCl = 4.6125×10^{-4} x 1000 / 25.00 = 0.01845 = **0.0185 mol dm^{-3}**
Question 9	$H_3PO_4 + 3\ KOH \rightarrow K_3PO_4 + 3\ H_2O$ Moles of H_3PO_4 = 0.500 x 25.00 / 1000 = 0.000125 moles Ratio 1:3 Moles KOH = 0.0125 x 3 = 0.0375 moles Concentration KOH = 0.0375 x1000 / 9.50 = 3.9474 = **3.95 mol dm^{-3}**
Question 10	$Na_2CO_3 + H_2SO_4 \rightarrow Na_2SO_4 + CO_2 + H_2O$ Moles of H_2SO_4 = 0.00500 x 20.00 / 1000 = 0.000100 moles Ratio 1:1 Moles Na_2CO_3 = 1.00×10^{-4} moles Concentration Na_2CO_3 = 1.00×10^{-4} x 1000 / 18.25 Concentration Na_2CO_3 = 0.005479 = **0.00548 mol dm^{-3}**
Question 11	$Ca(OH)_2 + 2\ HCl \rightarrow CaCl_2 + 2\ H_2O$ Moles of HCl = 0.0500 x 9.36 / 1000 = 4.68×10^{-4} moles Ratio 1:2 Moles HCl = 4.68×10^{-4} x ½ = 2.34×10^{-4} moles Concentration $Ca(OH)_2$ = 2.34×10^{-4} x 1000 / 10.00 a) Concentration $Ca(OH)_2$ = **0.0234 mol dm^{-3}** b) $M_r[Ca(OH)_2]$ = 40.1 + 2x17 = 74.1 Concentration / solubility = 0.0234 x 74.1 = 1.73394 = **1.73 g dm^{-3}**

Question 12

$K_2CO_3 \quad + \quad 2\,HCl \quad \rightarrow \quad 2\,KCl \quad + \quad CO_2 \quad + \quad H_2O$

Moles of HCl = 0.0500 x 13.50 / 1000 = 6.75 x 10^{-4} moles

Ratio 1:2

Moles K_2CO_3 = 6.75 x 10^{-4} x ½ = 3.375 x 10^{-4} moles in a 10.00 cm^3 portion

Moles in whole 100.0 cm^3 = 10 x 3.375 x 10^{-4}
= 3.375 x 10^{-3} moles

$M_r[K_2CO_3]$ = 2x39.1 + 60 = 138.2

Mass of K_2CO_3 = 3.375 x 10^{-3} x 138.2 = 0.466425 = **0.466 g**

Question 13

Expt	Ratio	Moles	Moles	Answer
A	1	0.0500 x 20.00 / 1000 = 0.00100	1 x 0.00100	0.00100 x 1000 / 13.65 = **0.0733 mol dm^{-3}**
B	2	0.00575 x 25.00 / 1000 = 1.4375 x 10^{-4}	2 x 1.4375 x 10^{-4} = 2.875 x 10^{-4}	2.875 x 10^{-4} x 1000 / 9.85 = **0.0292 mol dm^{-3}**
C	0.5	0.755 x 10.00 / 1000 = 0.00755	0.5 x 0.00755 = 0.003775	0.003775 x 1000 / 44.45 = **0.0849 mol dm^{-3}**
D	2	0.250 x 25.00 / 1000 = 0.00625	2 x 0.00625 = 0.0125	0.0125 x 1000 / 0.625 = **20.0 cm^3**
E	2	0.475 x 15.79 / 1000 = 0.00750025	2 x 0.00750025 = 0.0150005	0.0150005 x 1000 / 0.750 = **20.0 cm^3**

Question 14	a) Release of carbon monoxide b) Heating c) $M_r[SrSO_4]$ = 87.6 + 96.1 = 183.7 Moles strontium sulfate = 46.2 / 183.7 = 0.251497 moles $M_r[SrO]$ = 87.6 + 16 = 103.6 Balancing numbers on all strontium compounds is 1 making the reaction ratio is 1 : 1 : 1 : 1 Moles of strontium oxide = 12.5 / 103.6 = 0.120656 Yield = 100 x 0.120656 / 0.251497 = 47.975 % = **48.0 %** Alternative method: Maximum moles of SrO = 0.251497 moles Maximum mass = 0.251497 x 103.6 = 26.055 g Yield = 100 x 12.5 / 26.055 = 47.975 % = **48.0 %** d) Equation: $SrO + 2\ HCl \rightarrow SrCl_2 + H_2O$ Moles of HCl = 0.500 x 9.60 / 1000 = 0.00480 moles Moles of SrO = ½ x 0.00480 = 0.00240 moles Mass of SrO = 0.00240 x 103.6 = 0.24864 g Purity = 100 x 0.24864 / 0.256 = 97.125 = **97.1 %**
Question 15	Equation: $MHCO_3 + HCl \rightarrow MCl + CO_2 + H_2O$ Moles HCl = 0.150 x 20.21 / 1000 = 0.0030315 mol 1:1 reaction so moles of $MHCO_3$ = 0.0030315 mol $M_r[MHCO_3]$ = 0.255 / 0.0030315 = 84.1 The HCO_3 part has a M_r = 1 + 60 = 61 So $A_r[M]$ = 84.1 – 61 = **23.1 making M sodium**
Question 16	Equation: $M_2CO_3 + 2\ HCl \rightarrow 2\ MCl + CO_2 + H_2O$ Moles HCl = 0.200 x 25.33 / 1000 = 0.005066 mol 1:2 reaction so moles of M_2CO_3 = ½ x 0.005066 = 0.002533 mol $M_r[M_2CO_3]$ = 0.350 / 0.002533 = 138.2 The CO_3 part has a M_r = 60 So $A_r[M]$ = ½ x (138.2 – 60) = **39.1 making M potassium**

Question 17	Moles of KOH = 0.250 x 16.70 / 1000 = 0.004175 Moles of KOH required for whole flask = 0.004175 x 100.0 / 25 = 0.0167 mol Moles of H_3PO_4 added in step 1 = 0.05 x 25.00 / 1000 = 0.00125 Equation for phosphoric acid is H_3PO_4 + 3 KOH Moles of KOH neutralised by phosphoric acid = 0.00125 x 3 = 0.00375 mol Moles of KOH neutralised by ethanedioic acid = 0.0167 – 0.00375 = 0.01295 mol Ethanedioic acid is diprotic so requires 2 moles of KOH to neutralise it. Moles of ethanedioic acid = ½ x 0.01295 = 0.006475 mol in the whole flask $M_{r[(CO_2H)_2]}$ = (44 + 1) x 2 = 90 Mass of ethanedioic acid in whole flask = 0.58275 g The flask's ethanedioic acid was extracted from 50.0 g of rhubarb leaves. Mass of ethanedioic acid in 100 g of leaves = 1.1655 g = **1.17 g** (1.17 % by mass)
Question 18	Equation: $MCO_3 + 2\ HNO_3 \rightarrow M(NO_3)_2 + CO_2 + H_2O$ Moles HNO_3 = 0.250 x 15.75 / 1000 = 0.0039375 mol 1:2 reaction so moles of MCO_3 = ½ x 0.0039375 = 0.00196875 mol $M_{r[MCO_3]}$ = 0.166 / 0.00196875 = 84.3 The CO_3 part has a M_r = 60 So $A_{r[M]}$ = 84.3 – 60 = **24.3 making M magnesium**
Question 19	Equation: $M_2CO_3 + H_2SO_4 \rightarrow M_2SO_4 + CO_2 + H_2O$ Moles H_2SO_4 = 0.0500 x 23.04 / 1000 = 0.001152 mol 1:1 reaction so moles of M_2CO_3 = 0.001152 mol Moles of M_2CO_3 in whole flask = 0.001152 x 10 = 0.01152 mol dm^{-3} $M_{r[M_2CO_3]}$ = 0.850 / 0.01152 = 73.8 The CO_3 part has a M_r = 60 So $A_{r[M]}$ = ½ x (73.8 – 60) = 6.8924 = **6.9 making M lithium**

Question 20	
	a) Moles of NaOH = 0.0500 x 27.66 / 1000 = 0.001383 mol Reaction between HCl and NaOH is 1:1 so moles of HCl = 0.001383 mol Moles of HCl in whole flask = 0.001383 x 250 / 20 = 0.0172875 mol Summing the two reactions means the ratio of PCl_5 to HCl = 1 : 5 Moles of PCl_5 = 0.0172875 / 5 = 0.0034575 mol $M_r[PCl_5]$ = 31.0 + 5 x 35.5 = 208.5 Mass of PCl_5 in sample = 0.0034575 x 208.5 = 0.7209 g = **0.721 g** Purity = 100 x 0.721 / 0.875 = **82.4 %** b) Assumption is that there are no other reactions of NaOH with the impurity chemicals in the sample.

Examination Style Questions

Question 1	a)	Moles of HCl = 0.550 x 25.00 / 1000 = 0.01375 mol (1) Ratio 1:1 Moles NaOH = 0.01375 mol (1) Concentration NaOH = 0.01375 x 1000 / 18.55 = 0.74124 (1) Concentration to 3.s.f = **0.741 mol dm $^{-3}$** (1)
	b)	Colourless to (pale) pink (1)
Question 2	a)	Moles of H_2SO_4 = 0.850 x 20.00 / 1000 = 0.0170 mol (1) Ratio 1:2 Moles KOH = 2 x 0.0170 = 0.0340 mol (1) Concentration KOH = 0.0340 x 1000 / 37.25 = 0.91275 (1) Concentration to 3.s.f = **0.913 mol dm $^{-3}$** (1)
	b)	Red to (orange) or yellow (1)
Question 3	a)	Moles of $Ca(OH)_2$ = 0.0220 x 15.00 / 1000 = 3.30 x 10^{-4} mol (1) Ratio 2:1 Moles HCl = 2 x 3.30 x 10^{-4} = 6.60 x 10^{-4} mol (1) Concentration HCl = 6.60 x 10^{-4} x 1000 / 25.05 = 0.0263473 mol dm $^{-3}$ (1) Concentration HCl = **0.0263 mol dm $^{-3}$ to 3 s.f.** (1)
	b)	There is a limit to the solubility of Group 2 hydroxides in water. (1)
	c)	Strong acid reacting with a weaker alkali (1) This means the colour change at the endpoint is sharper with methyl orange than phenolphthalein. (1)

Question 4	a)	$3\ KOH_{(aq)} + H_3PO_{4(aq)} \rightarrow K_3PO_{4(aq)} + 3\ H_2O_{(l)}$ Left side including state symbols (1) Right side including state symbols (1)
	b)	Moles of H_3PO_4 = 1.25 x 10 / 1000 = **0.0125 mol** (1)
	c)	Ratio is 3 : 1 so moles of KOH = 3 x 0.0125 = **0.0375 mol** (1)
	d)	Conc. of the HCl = 0.0375 x 1000 / 43.25 = **0.867 mol dm^{-3} to 3 s.f.** (1)
Question 5	a)	Equation: $H_2SO_4 + Ca(OH)_2 \rightarrow CaSO_4 + 2\ H_2O$ Moles H_2SO_4 = 0.00500 x 25.00 / 1000 = 1.25 x 10^{-4} mol (1) 1:1 reaction so moles of $Ca(OH)_2$ = 1.25 x 10^{-4} mol (1) Concentration of $Ca(OH)_2$ = 1.25 x 10^{-4} x 1000 / 22.95 = 5.44662 x 10^{-3} (1) Concentration of $Ca(OH)_2$ = **0.00545 mol dm^{-3} to 3 s.f** (1) *(Answer is 3 s.f due to 1.50 value despite the others being 4 s.f.)*
Question 6	a)	Moles of HCl = 1.50 x 25.00 / 1000 = 0.0375 mol (1) Equation: $NaOH + HCl \rightarrow NaCl + H_2O$ Ratio = 1 : 1 Moles of NaOH = 0.0375 mol (1) Expected titre = 0.0375 x 1000 / 0.9000 = **41.7 cm^3 to 3 s.f** (1) (Answer is 3 s.f due to 1.50 value despite the others being 4 s.f.)
	b)	Colourless to (pale) pink (1)
Question 7	a)	Moles of H_2SO_4 = 1.25 x 15.00 / 1000 = 0.01875 mol (1) Equation: $2\ KOH + H_2SO_4 \rightarrow K_2SO_4 + 2\ H_2O$ Ratio = 2 : 1 Moles of KOH = 0.0375 mol (1) Expected titre = 0.0375 x 1000 / 2.25 = **16.7 cm^3 to 3 s.f** (1)
	b)	Strong acid and strong alkali so either indicator will work. (1)

Question 8	a)	Moles of $Ca(OH)_2$ = 0.00375 x 40.00 / 1000 = 0.000150 mol (1) Equation: $Ca(OH)_2 + 2\ HCl \rightarrow CaCl_2 + 2\ H_2O$ Ratio = 1 : 2 Moles of HCl = 0.000300 mol (1) Expected titre = 0.000300 x 1000 / 0.0850 = **3.53 cm^3 to 3 s.f** (1)
Question 9	a)	Moles of phosphoric acid = 0.01420 mol Equation: $3\ KOH + H_3PO_4 \rightarrow K_3PO_4 + 3\ H_2O$ (1) Moles of KOH = 3 x 0.01420 = 0.0426 moles (1) Titre = 0.0426 x 1000 / 0.950 = 44.842 cm^3 = **44.8 cm^3** (1)
	b)	Volume = 0.0142 x 1000 / 0.710 = **20.0 cm^3**
Question 10	a)	Moles of $Ca(OH)_2$ = 0.000685 x 100.0 / 1000 = 6.85×10^{-5} mol (1) Equation: $Ca(OH)_2 + H_2SO_4 \rightarrow CaSO_4 + 2\ H_2O$ Ratio = 1 : 1 Moles of H_2SO_4 = 6.85×10^{-5} mol (1) Expected titre = 6.85×10^{-5} x 1000 / 0.0250 = **2.74 cm^3 to 3 s.f** (1)
	b)	Ratio of concentrations is 1 : 5 (0.0250 / 0.00500 = 5) 5x lower in concentration so titre is 5 times larger = 2.74 x 5 = **13.7 cm^3** (1)
	c)	Titre is larger so the percentage uncertainty calculation has a larger denominator (1) 100 x 0.12 / 13.7 % rather than 100 x 0.12 / 2.74 % making the percentage uncertainty smaller. (1)

Question 11	a)	$NaOH_{(aq)}$ + $HCl_{(aq)}$ → $NaCl_{(aq)}$ + $H_2O_{(l)}$ Left hand side with state symbols (1) Right hand side with state symbols (1)
	b)	1 to 1 (1)
	c)	Titres with a set range, usually ± 0.2 cm^3 (1)
	d)	1, 3 and 4 (1) *[It does not have to be only two.]*
	e)	(22.85 + 22.85 + 22.95) / 3 = 22.8833 cm^3 (1)
	f)	Moles of HCl = 0.0500 x 22.8833 / 1000 = 0.001144 mol (1) Moles of NaOH = 0.001144 mol (1) Concentration of NaOH = 0.001144 x 1000 / 20.00 = 0.0572 mol dm^{-3} Concentration of NaOH = **0.0572 mol dm^{-3} to 3 s.f.** *[3 s.f. due to value of HCl concentration in the table.]*
Question 12	a)	500.0 cm^3 of 0.250 mol dm^{-3} $NaHCO_{3(aq)}$ needs 0.125 moles of $NaHCO_3$ (1) Mass of $NaHCO_3$ = 0.125 x (23.0 + 1.00 + 60.0) = 0.125 x 84.0 = 10.50 g (1) Dissolve the $NaHCO_3$ in about 200 cm^3 of distilled water then add to a 500.0 cm^3 volumetric flask. Rinse out the beaker into the flask with more distilled water. (1) Make up the flask to the 500.0 mark and mix. (1)
	b)	$NaHCO_3$ + HNO_3 → $NaNO_3$ + CO_2 + H_2O
	c)	$HCO_3^-{}_{(aq)}$ + $H^+{}_{(aq)}$ → $CO_{2(g)}$ + $H_2O_{(l)}$
	d)	Methyl orange (1) *[Due to hydrogen carbonate being a fairly weak alkali, phenolphthalein is less suitable.]* Yellow to red (1) To identify the exact volume when equal number of moles of nitric acid have been added to the sodium hydrogen carbonate. (1)
	e)	The reaction would fizz (1) *[Only asks for an observation not an identity of the gas.]*

Question 13

a) $2\ H_3PO_4 + 3\ Ca(OH)_2 \rightarrow Ca_3(PO_4)_2 + 3\ H_2O$

b)

Student 1	Percentage error	Student 2	Percentage error
Balance	100 x 0.005 x 2 / 0.37 = **2.70 %**	Balance	100 x 0.005 x 2 / 0.15 = **6.67 %**
Volumetric flask	100 x 0.3 / 250 = **0.12 %**	Volumetric flask	100 x 0.2 / 100 = **0.20 %**
Pipette	100 x 0.06 / 25 = **0.24 %**	Pipette	100 x 0.03 / 10 = **0.30 %**
Burette	100 x 0.02 x 2 / 17.0 = **0.24 %**	Burette	100 x 0.05 x 2 / 6.8 = **1.47 %**
Total % Uncertainty	**3.30 %**	**Total % Uncertainty**	± 8.64 %

Balance row (1), Volumetric flask row (1), Pipette row (1), Burette row (1), Student 1 Total % (1)

c) Student 1 (1)

Small mass on a 2 d.p. balance (1)

d)

	Student 1	Student 2
Total percentage uncertainty	**± 3.30 %**	**± 8.64 %**
Solubility of $Ca(OH)_2$	1.13 g dm^{-3}	1.13 g dm^{-3}
Uncertainty in g dm^{-3}	**1.13 ± 0.04** g dm^{-3} (1)	**1.13 ± 0.10** g dm^{-3} (1)
Range of values	1.09 to 1.17 g dm^{-3}	**1.03** to **1.23** g dm^{-3} (1)

e) Titration: increases the titre size as more calcium hydroxide added to flask from the sediment (1)

Solubility: increases the solubility from a titre that is falsely too high (1)

Question	Part	Answer
Question 14	a)	Moles of phosphoric acid added = 0.650 x 15.0 / 1000 = 0.00975 mol (1) Moles added to a titration = 0.000975 mol (tenth of flask) (1) H_3PO_4 + 3 KOH → K_3PO_4 + 3 H_2O Moles of KOH in titre = 0.350 x 18.87 / 1000 = 0.0066045 mol (1) Moles of KOH for the H_3PO_4 in a titration = 3 x 0.000975 = 0.002925 mol (1) Moles of KOH neutralised by tartaric acid = 0.0066045 – 0.002925 = 0.0036795 mol of KOH (1) Moles of tartaric acid = ½ x 0.0036795 = 0.00183975 moles in titration (1) Moles of tartaric acid in whole volumetric flask = 0.0183975 mol (1) M_r[tartaric acid] = 150 (stated in the question) Mass of tartaric acid = 0.0183975 x 150 = **2.76 g to 3 s.f.** (1)
	b)	100 x 2.76 / 175 = **1.58 %** by mass in tamarind pulp (1)
Question 15	a)	12.60 cm^3 and 12.70 cm^3 (1)
	b)	$Ca(OH)_{2(aq)}$ + 2 $HCl_{(aq)}$ → $CaCl_{2(aq)}$ + 2 $H_2O_{(l)}$ Left side including state symbols (1) Right side including state symbols (1)
	c)	Mean titre = 12.65 cm^3 Moles HCl = 0.0250 x 12.65 / 1000 = 3.1625 x 10^{-4} mol (1) Moles of $Ca(OH)_2$ = ½ x 3.1625 x 10^{-4} = 1.58125 x 10^{-4} mol (1) Concentration of $Ca(OH)_2$ = 1.58125 x 10^{-4} x 1000 / 25.00 = **0.006325 mol dm^{-3}** (1)
	d)	Systematic (1)
	e)	2.33 ± 0.005 g means **2.325 to 2.335 g** (1)
	f)	Burette uncertainty = 100 x 0.05 x 2 / 12.65 = **± 0.79 %** (1)
	g)	Total uncertainty = 0.79 + 0.24 + 0.10 + 0.43 = **± 1.56 %** (1) Range = ± 0.006325 x 1.56 / 100 = 0.00009867 = 0.00010 to 5 d.p. (1) Concentration of calcium hydroxide = **0.00633 ± 0.00010 mol dm^{-3}**

Question 16	a)	Moles of NaOH = 0.0100 x 28.02 / 1000 = 2.802×10^{-4} mol (1) Moles in volumetric flask = 2.802×10^{-4} x 250.0 / 20.00 = 0.0035025 mol (1) Reaction of HCl with NaOH is a 1:1 reaction Moles of HCl left after reaction of mineral = 0.0035025 mol Moles of HCl added to mineral = 1.00 x 200.0 / 1000 = 0.200 mol (1) Moles of HCl reacted with mineral = 0.200 – 0.0035025 = 0.1964975 mol (1) Moles of $Na_2Ca(CO_3)_2$ = ¼ x 0.1964975 = 0.0491244 mol (1) $M_r[Na_2Ca(CO_3)_2]$ = 2 x 23 + 40.1 + 2 x 60 = 206.1 (1) Mass of $Na_2Ca(CO_3)_2$ = 206.1 x 0.0491244 = 10.125 g (1) Percentage by mass = 100 x 10.1244 / 45.0 = **22.5 % to 3 s.f.** (1)
	b)	Titres are halved in size (1) This doubles the uncertainty in the titre values. (1)
Question 17	a)	Reaction 2: $Mg(OH)_2$ + 2 HCl → $CaCl_2$ + 2 H_2O Correct products (1) Correct balancing (1)
	b)	$M_r[CaCO_3]$ = 40.1 + 60 = 100.1 Moles of $CaCO_3$ = 0.500 / 100.1 = 0.004995 mol (1) This will neutralise: 2 x 0.004995 moles = 0.00999 moles of HCl (1) $M_r[Mg(OH)_2]$ = 24.3 + 2 x 17 = 58.3 Moles of $Mg(OH)_2$ = 0.583 / 58.3 = 0.0100 mol (1) This will neutralise: 2 x 0.0100 moles = 0.0200 moles of HCl (1) Total moles of HCl neutralised = 0.00999 + 0.0200 = 0.02999 mol (1) Concentration of stomach acid = 0.02999 x 1000 / 60.75 = **0.494 mol dm^{-3}** (1) *[Answer is to 3 s.f due to the 3 s.f masses in the question.]*

Question 18	a)	$2\ H_3NSO_3 + CaCO_3 \rightarrow Ca(H_2NSO_3)_2 + CO_2 + H_2O$ All 5 formulae correct (1) Correct balancing (1)
	b)	Moles of NaOH = 0.250 x 11.25 / 1000 = 0.0028125 mol (1) Reaction ratio of sulfamic acid to sodium hydroxide is 1 : 1. Moles of sulfamic acid in whole volumetric flask = 0.028125 mol (1) M_r[sulfamic acid] = 3 + 14 + 32.1 + 3x16 = 97.1 (1) Mass of sulfamic acid = 0.028125 x 97.1 = **2.73 g to 3 s.f.** (1)
	c)	0.028125 mol of sulfamic acid reacts with ½ x 0.028125 = 0.0140625 mol (1) $M_r[CaCO_3]$ = 40.1 + 60 = 100.1 Mass of $CaCO_3$ removed = 100.1 x 0.0140625 = **1.41 g to 3 s.f.** (1)
Question 19	a)	Moles of KOH = 0.0850 x 8.10 / 1000 = 6.885 x 10^{-4} mol (1) Monoprotic so moles of benzoic acid = 6.885 x 10^{-4} mol Concentration of $C_7H_6O_2$ = 6.885 x 10^{-4} x 1000 / 25 = 0.02754 mol dm^{-3} (1) $M_r[C_7H_6O_2]$ = 7x12 + 6 + 2x16 = 122 (1) Mass of benzoic acid = 122 x 0.02754 = 3.36 g Solubility of benzoic acid = 3.36 g dm^{-3} (1)
	b)	Increase the titre volume (1) Use a lower concentration of potassium hydroxide (1)
	c)	Lower temperature: **lower solubility** [some crystallises out] (1) Higher temperature: **same solubility** (1) [as no solid benzoic acid left to dissolve at the higher temperature]
	d)	Not disturb or pick up solid benzoic acid from the bottom of the flask. (1) Would increase the solubility by adding solid benzoic acid to the titration. (1)

Question 20

a)

Moles of NaOH = 0.0350 x 39.56 / 1000 = 0.0013846 mol (1)

Monoprotic so reaction ratio is 1 to 1 making moles of Substance B the same

Moles of Substance B = 0.0013846 mol in 15.00 cm^3 (1)

Moles in whole flask = 0.0013846 x 250 / 15.00
= 0.0230767 mol (1)

M_r = 3.785 / 0.0230767 = **164.0** (1)

b)

164 x 2 = **328** (1)

[If substance B was diprotic, there would be half the number of moles which will double the M_r.]

c)

Mass burned	**Carbon**	**Hydrogen**	**Oxygen**
1.955 g	5.245 g of CO_2	1.287 g of H_2	
All 3 in this row (1)	Mass of C = 5.245 x 12 / 44 = 1.4305 g	Mass of H = 1.287 x 2 / 18 = 0.143 g	Mass O = 1.955 – 1.4305 – 0.143 = 0.3815 g
All 3 in this row (1)	Moles C = 1.4305 / 12 = 0.11921 mol	Moles H = 0.143 / 1 = 0.143 mol	Moles O = 0.3815 / 16 = 0.02384 mol
Divide by smallest: (1)	0.11921 / 0.02384 = 5	0.143 / 0.02384 = 6	02384/ 0.02384 = 1
Empirical formula:	**C_5H_6O** (1) (M_r = 82)	Molecular formula:	**$C_{10}H_{12}O_2$** (1) (M_r = 164)

If you do not understand how to do this analysis then you will need Book 3 in this series; Analysis and Purity Calculations.

Question 21		
	a)	$MO + 2HNO_3 \rightarrow M(NO_3)_2 + H_2O$ (1)
	b)	1 to 1 (1)
	c)	*[This is a back titration.]* Moles of NaOH = 0.0400 x 41.87 / 1000 = 1.6748×10^{-3} (1) Moles of HNO_3 in 25.00 cm^3 = 1.6748×10^{-3} Moles of HNO_3 left in volumetric flask = 1.6748×10^{-3} x 500 / 25 = 0.033496 mol (1) Moles of HNO_3 added to metal oxide = 2.00 x 100 / 1000 = 0.200 mol (1) Moles of HNO_3 that reacted with the metal oxide = 0.200 – 0.033496 = 0.166504 mol (1) Moles of MO = ½ x 0.166504 = 0.083252 mol (1) [Equation in part a)] M_r[MO] = 3.355 / 0.083252 = 40.3 (1) 40.3 – 16 = 24.3 making the metal **magnesium** (1)
	d)	Titre volume would be 0.1 / 0.04 = 2.5 times smaller as concentration is 2.5 times larger. Original uncertainty = 100 x 0.06 x 2 / 41.87 = 0.287 % (1) New titre = 41.87 / 2.5 = 16.75 cm^3 New uncertainty = 100 x 0.06 x 2 / 16.75 = 0.716 % (1) 2.5 times greater uncertainty as 2.5 times smaller titre (1)

<table>
<tr><td rowspan="4">Question 22</td><td>a)</td><td>Moles of NaOH = 0.0500 x 33.52 / 1000 = 0.001676 mol (1)
Moles of $H_2C_2O_4$ = ½ x 0.001676 = 0.000838 mol (1)
Moles in whole 200.0 cm^3 of flask = 0.00838 mol (1)
M_r = 1.056 / 0.00838 = 126.0 (1)
$M_r[H_2C_2O_4]$ = 2 + 24 + 64 = 90.0 (1)
Water = 126 – 90 = 36 which is 2 x H_2O (1)
[The formula is $H_2C_2O_4.\mathbf{2}H_2O$]</td></tr>
<tr><td>b)</td><td>Ethanedioic acid is a weak acid, making methyl orange have an imprecise endpoint. (1)
Phenolphthalein changes colour above pH 7 so there will be a sharp endpoint (1)</td></tr>
<tr><td>c)</td><td>Pipette: 100 x 0.05 / 20.00 = 0.250 % (1)
Burette: 100 x 2 x 0.06 / 33.52 = 0.358 % (1)
Vol Flask: 100 x 0.25 / 200.0 = 0.125 % (1)
Balance: 100 x 2 x 0.0005 / 1.056 = 0.095 % (1)</td></tr>
<tr><td>d)</td><td>$M_r[H_2C_2O_4]$ = 2 + 24 + 64 = 90
Moles of ethanedioic acid = 1.056 / 90 = 0.011733 mol in 200 cm^3 (1)
Moles in titration = 0.0011733 mol (A tenth of the flask used.)
Diprotic acid so moles of NaOH = 2 x 0.0011733 = 0.023466 mol (1)
Titre = 0.023466 x 1000 / 0.0500 = 46.94 cm^3 (1)
[Alternative calculation: 33.52 x 126 / 90 = 46.93 cm^3]</td></tr>
</table>

Question 23		
	a)	$MnCO_{3(s)}$ + 2 $HCl_{(aq)}$ → $MnCl_{2(aq)}$ + $CO_{2(g)}$ + $H_2O_{(l)}$ Left side with state symbols (1) Right side with state symbols (1)
	b)	*[This is a back titration]* Moles of NaOH = 0.100 x 13.03 / 1000 = 0.001303 mol (1) NaOH : HCl is 1 to 1 making moles of HCl in the titration 0.001303 mol Moles of HCl left in the whole volumetric flask = 0.01303 mol (10 times value above for whole flask) (1) Moles of HCl added to mineral = 0.250 x 100 / 1000 = 0.0250 mol (1) Moles reacted with mineral = 0.0250 – 0.01303 = 0.01197 mol (1) Moles of $MnCO_3$ = 0.01197 / 2 = 0.005985 mol (1) $M_r[MnCO_3]$ = 54.9 + 60 = 114.9 Mass of $MnCO_3$ = 114.9 x 0.005985 = 0.6877 g (1) Percentage by mass = 100 x 0.6877 / 25.00 = **2.75 % to 3 s.f.** (1)
	c)	No other reactions with the $HCl_{(aq)}$ (1)
	d)	Less HCl in the volumetric flask meaning smaller titre (1) Smaller titre means more HCl appering to have reacted with the $MnCO_3$ making the percentage by mass appear higher (1)

Question 24		
	a)	*[This is a back titration]* Moles of KOH = 0.0505 x 31.93 / 1000 = 0.0016125 mol (1) KOH : HNO_3 is 1 to 1 making moles of HNO_3 in the titration 0.0016125 mol Moles of HNO_3 left in the whole volumetric flask = 0.016125 mol (x10) (1) Moles of HNO_3 added to hydroxide = 1.00 x 50 / 1000 = 0.0500 mol (1) Moles reacted with hydroxide = 0.0500 – 0.016125 = 0.033875 mol (1) Reaction: $M(OH)_2 + 2\ HNO_3 \rightarrow M(NO_3)_2 + 2\ H_2O$ (1) Moles of $M(OH)_2$ = 0.033875 / 2 = 0.0169375 mol (1) M_r of $(OH)_2.H_2O$ part = 2 x 17 + 18 = 52 Mass of $(OH)_2.H_2O$ = 52 x 0. 0169375 = 0.88075 g (1) Mass of M = 1.560 – 0.88075 = 0.67925 g (1) A_r[M] = 0.67925 / 0.0169375 = 40.1 (1) The element with an A_r of 40.1 = **calcium** (1)
	b)	M_rNi(OH)$_2$] = 58.7 + 34 = 92.7 Moles of nickel hydroxide = 1.545 / 92.7 = 0.01667 moles (1) Reaction ratio is 1: 2 [$Ni(OH)_2$ to HNO_3] Moles of HNO_3 to react with the $Ni(OH)_2$ = 2 x 0.01667 = 0.03333 moles (1) Moles of HNO_3 added = 0.0500 (as in part a) Moles of HNO_3 left in volumetric flask = 0.0500 – 0.03333 = 0.01667 (1) Moles of HNO_3 in the titration = 0.001667 = moles of KOH (1) Titre = 0.001667 x 1000 / 0.0505 = **33.0 cm^3**
	c)	$Ni(OH)_2$ is a transition metal so the nickel nitrate formed in solution will be coloured, masking the pale pink colour (1)

Question 25

a)

[This is a back titration]

Moles of NaOH = 0.750 x 4.68 / 1000 = 0.00351 mol (1)

NaOH : HNO_3 is 1 to 1 making moles of HNO_3 in the titration = 0.00351 mol

Moles of HNO_3 left in the whole volumetric flask = 0.00351 x 250 / 10.00 = 0.08775 mol (1)

Moles of HNO_3 added to $Fe(OH)_x$ = 2.00 x 150 / 1000 = 0.300 mol (1)

Moles reacted with $Fe(OH)_x$ = 0.300 – 0.08775 = 0.21225 mol (1)

If $Fe(OH)_2$ with M_r = 89.9	If $Fe(OH)_3$ with M_r = 106.9
Reaction ratio 1: 2 Moles of $Fe(OH)_2$ = 0.21225 / 2 = 0.106125 mol M_r = 9.540 / 0.106125 = 89.9 (1) Correct M_r	Reaction ratio 1: 3 Moles of $Fe(OH)_3$ = 0.21225 / 3 = 0.07075 mol M_r = 9.540 / 0.07075 = 134.8 (1) Incorrect M_r

The identity was iron(II) hydroxide with the formula $Fe(OH)_2$ making **x = 2** (1)

b)

More moles of hydroxide ions to react with the nitric acid (1)

This reduces the number of moles of nitric acid to titrate (1)

There appears to be less moles of iron hydroxide causing the M_r to be below 89.9 and therefore an unfeasible formula (1)

Explanation:

If sample was all $Fe(OH)_2$

Moles of $Fe(OH)_2$ = 9.54 g / 89.9 = 0.106 moles

Moles of OH^- ions = 2 x 0.106 = 0.212 mol

If the sample was 50% of each, then...

Moles $Fe(OH)_2$ = 4.77 g / 89.9 = 0.0531 moles

Moles $Fe(OH)_3$ = 4.77 g / 106.9 = 0.0446 moles

Moles of OH^- ions = 2 x 0.0531 + 3 x 0.0446 = 0.240 mol
(more OH^- ions to react with HNO_3)

Therefore the titre would decrease in size with number of moles of HNO_3 left to titrate being reduced.

Result would be less moles in the calculation of M_r causing a value below 89.9 which would be an unfeasible formula for iron hydroxide as Fe usually forms Fe^{2+} and Fe^{3+} ions.

c)

Reduce the concentration of the sodium hydroxide used in the titration to increase the titre size (1)

Question 26

a)

(i) Oxidation numbers:

FeO is +2, Fe_2O_3 is +3 (1) [both required for mark]

Fe_3O_4 is + $2^2/_3$ (1) [8 ÷ 3]

(ii) Mixed oxidation state with 1x Fe(II) and 2x Fe(III)

b)

[This is a back titration]

Moles of KOH = 2.00 x 34.39 / 1000 = 0.06878 mol (1)

KOH : H_2SO_4 is 2 to 1 making moles of H_2SO_4 in the titration 0.03439 mol (1)

Moles of H_2SO_4 left in the whole volumetric flask
= 0.03439 x 500 / 15.00 = 1.14633 mol (1)

Moles of H_2SO_4 added to the iron oxide
= 5.00 x 325 / 1000 = 1.625 mol (1)

Moles reacted with iron oxide
= 1.625 – 1.14633 = 0.47867 mol (1)

If FeO with M_r = 71.9	If Fe_2O_3 with M_r = 159.8	If Fe_3O_4 with M_r = 231.7
Reaction ratio 1: 1 with H_2SO_4 Moles of FeO = 0.4787 mol M_r = 25.50 / 0. 4787 = 53.3 (1) Incorrect M_r	Reaction ratio 1: 3 with H_2SO_4 Moles of Fe_2O_3 = 0.47867 ÷ 3 = 0.15956 mol M_r = 25.50 / 0.15956 = 159.8 **Correct M_r**	Reaction ratio 1: 4 with H_2SO_4 Moles of Fe_3O_4 = 0.47867 ÷ 4 = 0.11967 mol M_r = 25.50 / 0.11967 = 213.1 (1) Incorrect M_r

The identity was iron(II) hydroxide with the formula **Fe_2O_3** (1)

Acknowledgements

I hope you have found this book to contain useful and thought-provoking practise in addition to your existing questions and that it has enhanced your understanding. No doubt there may be unintentional mistypes and errors so any feedback to mrevbooks@outlook.com will be very much appreciated along with any requests for future question topics.

This book came out of the preparation of materials for teaching A-level Chemistry and the lack of practice questions in published whole course textbooks.

The questions originated as a range of homework tasks and tests. I therefore have to thank all the students who studied GCE Chemistry at both Uplands Community College and Beacon Academy in East Sussex where many of the questions were trialled and tested along with the other schools where I have taught in recent years and created materials.

I am also developing a series of videos with follow-up questions and answers supporting A-level course content. Search for MrEV Chemistry on YouTube. Please subscribe!

The biggest thanks of all must go to my partner Hayley who uses her not inconsiderable proof-reading and language skills to rectify the typing errors and poorly worded questions on the first drafts of the content and encouraged me to create these books in the first place.

Thank you for buying and I wish you every success in your future endeavours, wherever they may lead you.

Printed in Great Britain
by Amazon

48009024R00043